Full STEAM Ahead!

Engineering Everywhere

Many Kinds of Engineers

Robin Johnson

CRABTREE
PUBLISHING COMPANY
WWW.CRABTREEBOOKS.COM

Title-Specific Learning Objectives:

Readers will:

- Describe different types of engineers.
- Explain how engineers use math, science, and creative thinking to solve problems.
- Identify details that support the idea that engineers in many fields solve different problems.

High-frequency words (grade one)	**Academic vocabulary**
are, can, did, how, from, make, they, this	design, engineer, environment, field, fuel, machine, material, technology

Before, During, and After Reading Prompts:

Activate Prior Knowledge and Make Predictions:
Spread a sheet or blanket flat on the ground. Ask for enough volunteers to come stand on it, filling all but about one-quarter of the space. Challenge them to flip the sheet or blanket over so they are standing on the other side. But, they may not step off the sheet or touch the ground. Have all children suggest solutions.

When they are done, point out that it took some creative thinking to solve the problem. Hold up the book and read the title out loud. Explain that engineers are people who use creative thinking to solve problems. Ask them to say what they think the book will be about.

During Reading:
After reading pages 10–11, ask children to restate the kind of work the engineers described on those pages do. Then have them list details that help explain that idea.

After Reading:
Have children work in small groups to list as many different types of engineers that they can remember. Afterward, make a master list on the board, using simple language such as space engineers, ship engineers, and community engineers. For each type, ask children to describe the kinds of jobs each engineer does.

Author: Robin Johnson

Series Development: Reagan Miller

Editors: Bonnie Dobkin, Janine Deschenes

Proofreader: Melissa Boyce

STEAM Notes for Educators: Bonnie Dobkin

Guided Reading Leveling: Publishing Solutions Group

Cover, Interior Design, and Prepress: Samara Parent

Photo research: Robin Johnson

Production coordinator: Katherine Berti

Photographs:
Alamy: NG Images: p. 6
iStock: Opla: title page, p. 18; FatCamera: p. 5 (b); Zuraisham Salleh: p. 7 (b); SeventyFour: p. 11 (t); Zephyr18: p. 13 (t); energyy: p. 15 (t); Hispanolistic: p. 15 (b); monkeybusinessimages: p. 16; Tashi-Delek: p. 17 (t); andresr: p. 19 (t), p. 20; SerrNovik: p. 19 (b)

All other photographs by Shutterstock

Library and Archives Canada Cataloguing in Publication

Title: Many kinds of engineers / Robin Johnson.
Names: Johnson, Robin (Robin R.), author.
Description: Series statement: Full STEAM ahead! | Includes index.
Identifiers: Canadiana (print) 20200165895 | Canadiana (ebook) 20200165909 | ISBN 9780778771890 (hardcover) | ISBN 9780778772675 (softcover) | ISBN 9781427124609 (HTML)
Subjects: LCSH: Engineering—Juvenile literature. | LCSH: Engineers—Juvenile literature. | LCSH: Engineering design—Juvenile literature.
Classification: LCC TA149 .J655 2020 | DDC j620—dc23

Library of Congress Cataloging-in-Publication Data

CIP available at the Library of Congress
LCCN: 2019058913

Printed in Canada/112022/CPC20221114

Table of Contents

Looking for Answers.................... 4
Who Are Engineers?................ 6
Different Jobs 8
Mixing Materials 10
Mighty Machines..... 12
Using Electricity 14
In the Community.... 16
For the Environment 18
Working Together ... 20
Words to Know....... 22
Index and About the Author.... 23
Crabtree Plus Digital Code............ 23
STEAM Notes for Educators................ 24

Crabtree Publishing Company
www.crabtreebooks.com 1-800-387-7650
 In Canada: We acknowledge the financial support of the Government of Canada through the Canada Book Fund for our publishing activities.

Published in Canada
Crabtree Publishing
616 Welland Ave.
St. Catharines, Ontario
L2M 5V6

Published in the United States
Crabtree Publishing
PMB 59051
350 Fifth Avenue, 59th Floor
New York, New York 10118

Published in the United Kingdom
Crabtree Publishing
Maritime House
Basin Road North, Hove
BN41 1WR

Published in Australia
Crabtree Publishing
Unit 3 – 5 Currumbin Court
Capalaba
QLD 4157

Looking for Answers

Did you ever wonder how we get clean water in our homes? Do you know how we can travel from place to place quickly and safely? Engineers have the answers!

How can **technology** give some people extra help? Engineers keep finding new ways.

How does an airplane stay in the air? Ask an engineer!

How can we find and share information? Engineers found the **solutions**!

Who Are Engineers?

Engineers are people who use math, science, and **creative thinking** to solve problems. First, they figure out what is needed. Then, they **design** and build things to meet the need.

The machines and tools engineers make are called technology.

Some engineers design robots. A robot is a type of technology. It does tasks that are hard for people to do.

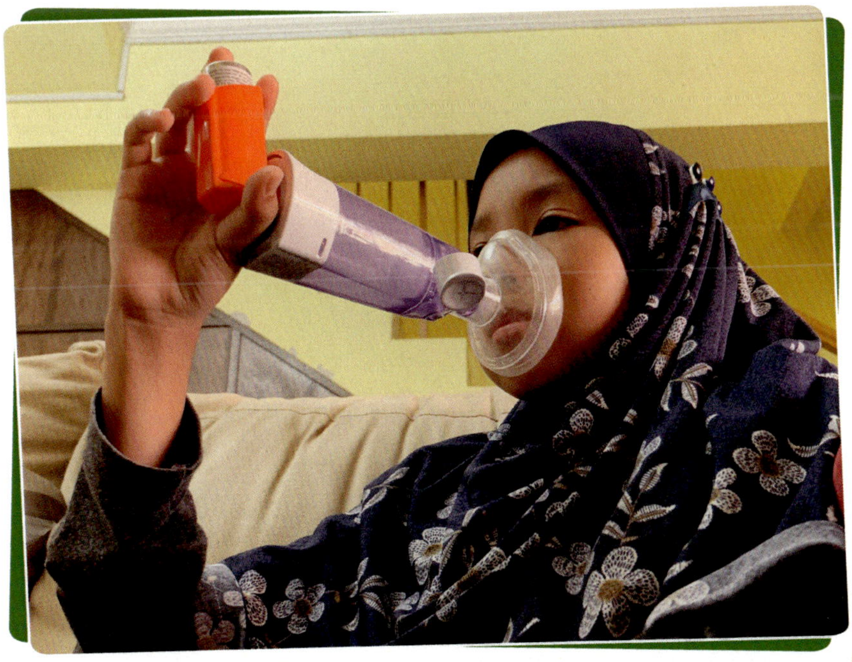

Some engineers design technology to solve health problems. This girl is using a machine that helps her take medicine.

Different Jobs

There are many kinds of engineers. They work in different areas, called **fields**. Engineers in different fields solve different types of problems.

Some engineers make technology that helps farmers grow crops.

Some engineers design boats and ships to help travel on water. They also **inspect** ships to make sure there are no problems.

Engineers also work to help the environment. This engineer makes sure people can get the wood they need without harming the forest.

Mixing Materials

Some engineers design ways to change **raw materials** into useful items. They make food, medicine, clothing, paper, and other things we need. They also design ways to make the items quickly and safely.

Some engineers in this field make new and better medicines.

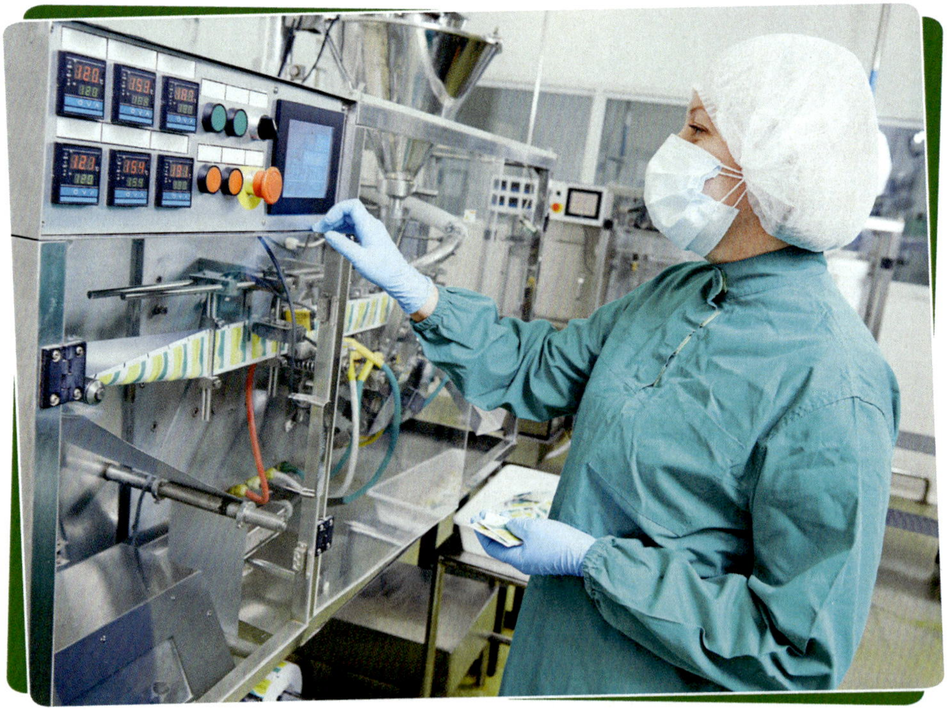

Engineers work with food to make it healthier. This engineer is making soy milk.

Engineers design useful materials like paper and cloth. They also design new materials, such as plastic that is easily recycled.

Mighty Machines

Engineers in another field design machines, tools, and other objects. These technologies help people do work. They also make our lives easier and safer.

This engineer is designing a new part for a car.

Some engineers in this field design machines used in buildings. This engineer is checking the strong metal ropes that help an elevator move up and down.

This engineer is inspecting a crane. A crane is a machine that lifts heavy loads.

Using Electricity

Some kinds of technology run on electricity. Engineers in another field design these technologies. They also design ways to bring electricity to homes and other buildings.

Electricity is a type of energy. We use it to light our homes. It also gives power to technologies such as computers.

This engineer works at a power station. He makes sure that electricity travels on wires to people's homes.

An engineer in this field might design a light bulb that uses less energy.

In the Community

Some engineers design things that make communities work well. They design buildings, roads, and bridges. They design ways to bring clean water into homes and take dirty water out.

Engineers in this field design bridges so we can cross over water.

Engineers make careful plans to design strong, safe buildings. The buildings give people places to live and work.

Engineers in this field design ways to bring clean water to people. We drink the water. We use it for washing and cooking.

For the Environment

Some engineers work to protect the environment. They look for ways to control **pollution**. They design technology that meets people's needs without harming the planet.

This engineer designs ways to clean the water that comes from factories.

Engineers in this field also find ways to reduce waste.

These students are learning how engineers use wind to power our homes.

Working Together

Engineers from different fields work together to solve problems. They share what they know with each other. Together the engineers come up with many ideas and solutions.

Many kinds of engineers are needed to help build a shopping mall. The huge building must be strong and safe. It needs electricity, water, and heat. There are escalators and elevators that carry people up and down.

What do you think these engineers are designing? What problem could it solve?

Many kinds of engineers are needed to design a car. Cars have engines and other moving parts. They have lights and devices that use electricity. They need fuel to run. They are made of metal, rubber, and many other materials.

Words to Know

creative thinking [kree-EY-tiv THING-king] noun Using your mind to make up new and original ideas

design [dih-ZAHYN] verb To make a plan for how something is made or built

fields [feeldz] noun Areas or subjects of work or interest

inspect [in-SPEKT] verb To look closely at something

pollution [p*uh*-LOO-sh*uh*n] noun Waste that makes the environment unclean

raw materials [raw m*uh*-TEER-ee-*uh*ls] noun Things from which something is made

solutions [s*uh*-LOO-sh*uh*ns] noun Answers to a problem

technology [tek-NOL-*uh*-jee] noun Tools that make life easier, safer, and more fun

A noun is a person, place, or thing.

A verb is an action word that tells you what someone or something does.

An adjective is a word that tells you what something is like.

Index

bridges 16
buildings 16–17
electricity 14–15, 20–21
environment 9, 18–19
farming 8
health 7
machines 6–7, 12–13
materials 10–11, 21
roads 16
water 4, 9, 16–18, 20

About the Author

Robin Johnson is a freelance author and editor who has written more than 80 children's books. When she isn't working, Robin builds castles in the sky with her engineer husband and their two best creations—sons Jeremy and Drew.

To explore and learn more, enter the code at the Crabtree Plus website below.

www.crabtreeplus.com/fullsteamahead

Your code is:
fsa20

STEAM Notes for Educators

Full STEAM Ahead is a literacy series that helps readers build vocabulary, fluency, and comprehension while learning about big ideas in STEAM subjects. *Many Kinds of Engineers* helps readers learn about different types of engineers and identify details that help them understand each one. The STEAM activity below helps readers extend the ideas in the book to build their skills in engineering and arts.

Card Challenge

Children will be able to:

- Explain an engineer's design process.
- Analyze a problem.
- Use creative thinking to solve the problem.

Materials

- Engineering Design Process Worksheet
- Card Challenge Handout
- For each group: at least 30 index or playing cards, safety scissors, tape, rubber bands, glue, and one small piece of cardboard or poster board to use as a base

Guiding Prompts

After reading *Many Kinds of Engineers*, ask children:

- What does an engineer do?
- Describe some types of engineers you read about. What does each do?
- How do engineers solve problems?

Activity Prompts

Ask children to summarize the job of an engineer. Tell them that today they will become engineers and do some creative problem solving!

Distribute the Engineering Design Process Worksheet. Explain to children that these are the steps engineers go through when solving a problem. Read through the steps, asking for volunteers to restate each one in their own words.

Now hand out the Card Challenge Handout. Read the challenge aloud to students: "Your challenge is to make the tallest or strongest structure you can, using only cards and whatever you need to hold them together."

Divide students into groups of three or four, providing each with a set of materials. Have them follow the steps of the design process, keeping track of their ideas on the handout. After a set amount of time, see whose structure is the tallest. Then test for the strongest structure by putting a heavy object on top.

Extensions

- Ask each group to present their structure to the class. Have them explain their ideas, their reasons for choosing the solution they did, and any improvements they would make.

To view and download the worksheet, visit **www.crabtreebooks.com/resources/printables** or **www.crabtreeplus.com/fullsteamahead** and enter the code **fsa20**.